PENGUIN BOOKS

the Language of Flowers

the Language of Flowers

A treasury of traditional meanings

PENGUIN BOOKS

PENGUIN BOOKS

Published by the Penguin Group
Penguin Group (Australia)
250 Camberwell Road, Camberwell, Victoria 3124, Australia
(a division of Pearson Australia Group Pty Ltd)

New York   Toronto   London   Dublin   New Delhi
Auckland   Johannesburg

Penguin Books Ltd, Registered Offices: 80 Strand, London, WC2R 0RL, England

First published by Penguin Group (Australia), a division of Pearson Australia
Group Pty Ltd, 2012

10 9 8 7 6 5 4 3 2 1

Cover and text design by Marley Flory © Penguin Group (Australia)
All illustrations courtesy of Sherardian Library of Plant Taxonomy,
One of the Bodleian Libraries of the University of Oxford
Text by Sue Behrent
Typeset in Neutraface Text
Scanning and separations by Splitting Image P/L, Clayton, Victoria
Printed in China by 1010 Printing International Limited

National Library of Australia
Cataloguing-in-Publication data:

The Language of Flowers
ISBN: 9780143206231 (hbk.)
Includes Index.
1. Flower language

398.368213

penguin.com.au

Contents

Introduction

The ascription of particular properties to species of plant and flower has a long tradition. In the Middle Ages, certain herbs were believed to have magical powers, while Renaissance art often depicted saints with flowers symbolic of their moral virtues. At various points in history, the Assyrians, Chinese, Indians and Egyptians – amongst others – all developed their own dictionary of flowers conveying certain meanings.

However, the concept as we recognise it today was derived from the Turks. In the early eighteenth century, Lady Mary Wortley Montagu, wife of the British Ambassador to Turkey (and travel writer and feminist), wrote from Constantinople to friends detailing the variety of messages the Turks exchanged via the medium of flowers.

The notion caught on, and throughout the Victorian era of the eighteenth and nineteenth century the subtle language of flowers – or floriography – became an important and popular part of refined society across Europe.

A vast nuanced vocabulary developed which circumvented the strict social mores of the time, with coded communications transmitted through the variety, size, arrangement and even the scent of particular plants. Lovers could declare their ardour, rivals could promise revenge, the cad could show his disdain and the devoted his constancy.

While floriography was at its zenith during the Victorian period, the language of flowers still captures the imagination today.

Fun, Frivolity
& Assorted
Engaging Pastimes

Activity Thyme

Amusement, frivolity Bladder Nut Tree

Anticipation Gooseberry

Banter, merriment and jest

  Southernwood

Christmas Poinsettia

Curiosity Sycamore

Dreaminess Poppy (variegated)

Dreams Osmunda

Eagerness Pelargonium

Thyme

London Pride

Expectation Zephyr Flower

Expected meeting Nutmeg Geranium

Fanciful Lupin (pink)

Festivity Parsley

Frivolity London Pride

Laugh at trouble Harlequin

Mirth Saffron Crocus

Music Reeds (with their panicles)

Philosophy Pitch Pine

Poetry Sweetbrier

Rendezvous Chickweed

Satire Prickly Pear

Splendour of spring Claytonia

Sporting, fun Foxtail Grass

Surprise Betony · Truffle

Unexpected meeting Lemon Geranium

Variety Aster

Verse Wild Vine

Wit Ragged Robin

Happiness

Always cheerful Coreopsis

   • Holly (variegated)

Always delightful, a star ever bright

   Cineraria

Always happy Lupin (white)

Brightness Ivy (variegated)

Cheerfulness Chrysanthemum

Content Fern Moss

Contentment Bluets

Gaiety Butterfly Orchid

Gladness Myrrh

Happiness Mugwort • Sultan (sweet)

Mugwort

Lesser Celandine

Happiness in old age American Starwort

Happiness, laughter Saffron Crocus

I'm too happy! Cape Jasmine

Joy Dahlia (red) · Persimmon · Wood Sorrel

Joys to come Lesser Celandine

Poor but happy Vernal Grass

Rural happiness Violet (yellow)

Fickleness, Gossip & Further Examples of Weak Conduct

Argument Fig

Aversion China Pink · Indian Pink (single)

Bad-tempered Crabapple Blossom

Bad temper, sour disposition Barberry

Changeable disposition Rye Grass

Criticism Cucumber

Difficulty Blackthorn

Diffidence Cyclamen · Primula

Disappointment Carolina Syringa

Crabapple Blossom

Columbine

Dissention Pride of China

Division, rupture of contract

   Straw (broken)

Fickleness Lady's Slipper · Larkspur (pink)

Folly Columbine

Foolishness Pomegranate

Gossip Cobaea

Gossip, rumour Grape Vine

Ill-timed wit Wild Sorrel

Impatience Balsam (yellow)

Burdock

Impatient of absence Corchorus

Imperfection Henbane · Heps and Haws

Importunity Burdock

Inconstancy Evening Primrose

Inconstancy in love Wild Honeysuckle

Indecision China Aster (single) · Wild Daisy

Indifference Candytuft · Mustard Seed

Candytuft

Buttercup

Indifference, apathy Pigeon Berry · Senvy

Indifference, coldness Vitex Agnus Castus

Indiscretion Reeds (split)

Indiscretion, docility Bulrush

Ingratitude Buttercup

Insinuation Great Bindweed

Intoxication Vine

Mistake Fly Orchid

Mistake, error Bee Orchid

Paternal error Cardamine

Presumption Snapdragon

Quarrel Corn (broken)

Reproach Rose (withered)

Rudeness Clotbur · Xanthium

Severity Thorns (branch of)

Shame Peony

Silliness Fool's Parsley · Geranium (scarlet)

Peony

Stupidity, indiscretion Almond

Submission Grass

Tardiness Flax-leaved Golden Locks

Uncertainty Apricot Blossom

Vulgar mind African Marigold

Weakness Moschatel · Musk Plant

You are cold Hortensia

You are hurting me! Elecampane

Your temper is too hasty Grammanthes

Money, Luck & Business

Austerity Common Thistle

Bad luck Clover (five-leaved)

Compensation Poor Robin's Plantain

Good luck Clover (four-leaved) · Toadstool

I offer you my fortune Caceolaria

Luxury Buckeye

Poverty Evergreen Clematis

Profit Cabbage

Prosperity Beech Tree

Beech Tree

Coriander

Providence Clover (purple)

Riches Corn · Wheat

Royalty Comet Orchid

Victory Palm

Wealth doesn't always lead to happiness
   Auricula

Worth (hidden) Coriander

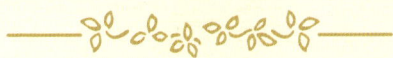

Purity, Modesty
& Youth

Blushes Marjoram

Charm and innocence Rose (white)

Charming simplicity Rose (wild)

Chastity Lily of the Valley

Docility Rush

Early childhood and youth Primrose

Freshness Damask Rose

Humility Small Bindweed · Ground Ivy

· Field Lilac

Ingenuous simplicity

Mouse-eared Chickweed

Innocence Daisy

Mouse-eared Chickweed

Broom

Innocence, naïveté Freesia

  · Rosebud (white)

Meekness Birch Tree

Mildness Mallow

Modesty Trillium · Sweet Violet

Modesty, innocence and purity

  Violet (white)

Neatness, humility Broom

Pure and guileless Verbena (white)

Pure and lovely Rosebud (red)

Purity Star of Bethlehem

Purity, chastity Orange Blossom

Purity, cleanliness Hyssop

Purity, delicacy, constancy Cornflower

Purity, modesty Arum Lily

Purity of heart Baby's Breath · Water Lily

Reward of chastity Roses (crown of)

Reward of virtue Roses (garland of)

Simplicity Rose (single)

Sweet disposition Lavatera

Lavatera

Sweet innocence Scilla (white)

Sweetness Sultan (white)

Timidity Four O'clock Flower

Timidity, splendid beauty Amaryllis

Virtue Mint

Youthful gladness Spring Crocus

Youthful innocence Lilac (white)

Youthful love Red Catchfly

Love's Many Stages

Admiration Amethyst

Adoration Dwarf Sunflower

Adulation Cacalia

Affection Morning Glory

  • Mossy Saxifrage • Sorrel

Alas! My poor heart! Carnation (red)

Ambassador of love Cabbage Rose

Ardour Cocoa Plant • Lady's Smock

Attachment Indian Jasmine

Lady's Smock

Motherwort

Clinging affection Irish Ivy

Concealed love Motherwort

Confession of love Moss Rosebud

  · Tulip (red)

Conjugal love Lime Tree

Cure for heartache Cranberry

  · Swallowwort

Devoted affection Honeysuckle

Devotion Night-scented Stock

  · Peruvian Heliotrope

Do not refuse me Carrot Flower

Do you still love me? Lilac (mauve)

Early attachment Thornless Rose

Enduring love Gorse

Esteem, but not love Spiderwort

Esteem, love, excellence Strawberry

Estranged love Lotus Flower

Filial love Virgin's Bower

First emotions of love Lilac (purple)

Forsaken Laburnum · Anemone

· Common Willow

Fraternal love Woodbine

Strawberry

Shepherd's Purse

High esteem Garden Sage

Hopeless love Tulip (yellow)

I am dazzled by your charms Ranunculus

I am hurt Mustard

I cling to thee Vetch

I dare you to love me Tiger Lily

I desire a return of affection Jonquil

I offer you my all Shepherd's Purse

I only dream of love Moonflower

I prefer you Geranium (pink)

If you love me you will find out

Maiden's Blush Rose

Immortality, unfading love

Globe Amaranth

Lasting love Chinese Primrose

Let me go Butterfly Weed

Love Chrysanthemum (red) · Myrtle

Love and desire Rose (red)

Love at first sight Plains Coreopsis

· Love-in-idleness

Love forsaken Creeping Willow

Love is dangerous Carolina Rose

Love of nature Magnolia

Love returned Ragwort

Ragwort

Rose (pink)

Love, sweet and secret Honey Flower

Love undiminished Flowering Dogwood

Love, beauty Cinnamon

Love, reverence Carnation

Love's oracle Moon Daisy

Maternal love and affection Cinquefoil

Matrimony American Linden

Motherly love Windflower

Only deserve my love Rose Campion

Our love is perfect happiness Rose (pink)

Our love will be fruitful Tea Rose

Passion Fleur-de-lis · Lemon Tree

Preference Apple Blossom

Pure and true love Carnation (white)

Reconciliation Filbert

Regard Daffodil (double)

Rejection, rebuff Balsam (rose)

Rupture Greek Valerian

Secret love Acacia (yellow) · Clove

Slighted love Chrysanthemum (yellow)

The heart's mystery Polyanthus (crimson)

Greek Valerian

Timorous love Apricot Blossom

Touch me not! Balsam (red)

True love Forget Me Not

Unfortunate love Scabiosa

Union Lancaster Rose

Woman's love, I'll never forget you

Carnation (pink)

You are charming Cluster Rose

You are perfect Pineapple

Your charms are engraved on my heart

Spindle Tree

Yours until death Yucca

Fidelity

Accept a faithful heart Weigela

Consistency Globe Flower

Constancy Chimney Bellflower · Bluebell

· Sunflower

Faithfulness Dog Violet

· Sweet Violet (blue)

Faithfulness, devotion Heliotrope

Fidelity Ivy

Ivy

Speedwell

Fidelity, allegiance Plum Tree · Veronica

Fidelity in adversity Wallflower

Fidelity in love Lemon Blossom

Fidelity of women Speedwell

Loyalty Shamrock

The Seven
Deadly Sins

Affectation Cockscomb

Anger Whin

Anger, wrath Fumitory

Boastfulness Hydrangea

Conceit Nettle Tree

Egotism Narcissus

Ennui Moss

Extravagance Poppy

Grandeur Ash Tree

Greed Auricula (scarlet)

Haughtiness Larkspur (purple)

Narcissus

Kingcup

Haughtiness, conceit Tall Sunflower

Hunger for riches Kingcup

Idleness Fig Marigold

· Mesembryanthemum

Jealousy French Marigold

Jealousy, diminishing of love

Rose (yellow)

Lowliness, envy, remorse Rubus

Pretension Spiked Willow-herb

Pride Hundred-leaved Rose

· Gloire de Santenay Rose · Amaryllis

Meadowsweet

Pride of riches Polyanthus

Rivalry Rocket

Self-seeking Clianthus

Temptation Apple · Quince

Uselessness Diosma · Meadowsweet

Vice Ray Grass (Darnel)

You can boast too much Stephanotis

Beauty

Always lovely Indian Pink (double)

Beautiful eyes Tulip (variegated)

Beauty Daisy (parti-coloured)

Beauty ever new Rose (monthly)

Beauty is your only attraction

  Rose (Japanese)

Beauty of the mind Clematis

  • Old Man's Beard

Capricious beauty Musk Rose

Delicate beauty Flower-of-an-hour

  • Hibiscus

Old Man's Beard

Stock

Feminine beauty Cherry-apple Blossom

Fresh beauty China Rose

Glorious beauty Glory Flower

Lasting beauty Stock

Lasting beauty, bonds of affection

Gillyflower

Magnificent beauty Calla Lily

Mature charms Cattleya

Neglected beauty Throatwort

Perfected loveliness

Japanese Camellia (white)

Cherry Blossom

Perfection of female loveliness Justicia

Spiritual beauty Cherry Blossom

Thou art all that is lovely Rose (Austrian)

Unconscious beauty Rose (burgundy)

You are a divine beauty Marsh Marigold

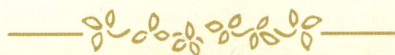

Grace & Elegance

Aster (detail)

Bewitching grace Lady's Tresses

Delicacy, grace Aster

Dignity Laurel-leaved Magnolia · Elm

Elegance Acacia (pink, red or white)

· Locust Tree

Aster

Cowslip

Elegance and fortitude Campion

Elegance and grace Yellow Jasmine

Elegance and strength Bamboo

Grace Rose (multiflora)

Mature elegance Pomegranate Flower

Winning grace Cowslip

Cruelty & Deception

Betrayal Judas Tree

Betrayed White Catchfly

Blackness, you are hard Ebony Tree

Cold-heartedness Lettuce

Coquettishness Daylily

Counterfeit Mock Orange

Cruelty Nettle

Deceit Venus Flytrap

Deceit, lies Dogbane

Deceitful charms Thorn Apple

Deception White Cherry Tree

　• Winter Cherry

Defamation Madder

Madder

Distrust Lavender

Falsehood Bugloss · Manchineel Tree

Falsehood, lies Lily (yellow)

Flattery Venus' Looking-glass

Insincerity Foxglove

Scandal Hellebore

Treachery Bilberry · Dahlia

Treachery, deceit Cherry Laurel

Treason Whortleberry

You are queen of the coquettes

   Dame's Rocket

You are spiteful Common Stinging Nettle

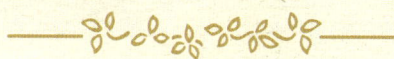

Fascination, Enchantment & Belief

Enchanter's Nightshade (detail)

Enchantment Holly Herb

· Rosebud (monthly) · Vervain

Fascination Fern

Imagination Lupine

Magic, sorcery Enchanter's Nightshade

Enchanter's Nightshade

Persuasion Common Hibiscus

Religious devotion Campion

Religious enthusiasm Schinus

Religious faith, superstition
 Passion Flower

Spellbound Witch-hazel

Tree of the fairies Alder Tree

Uneasiness Garden Marigold

Kindness, Benevolence & Other Acts of Friendship

A friend in need Adam's Needle

Acknowledgement Canterbury Bells

Advice Rhubarb

Agreement Corn Straw

Amiability White Jasmine

An accommodating disposition Valerian

Be my support Black Bryony

Benevolence Calycanthus · Potato

Charity Wild Grape · Turnip

Compassion Allspice · Elder

Black Bryony

White Mullein

Concord Lote Tree

Desire to please Mezereon

Do me justice Chestnut Tree

Early friendship Periwinkle (blue)

Family union Verbena (pink)

Forgive and forget Scilla (blue)

Forgiveness of injuries Cinnamon Tree

Friendship Acacia

Generosity Orange Tree

Good nature White Mullein

Good wishes Sweet Basil

Goodness Good King Henry · Mercury

Goodness, decency Goosefoot

Honesty, sincerity Honesty

Hospitality Oak Tree

I am not changed Pyrethrum

I forgive you Nemophila

I have a message for you Iris

Keep this for my sake Showy-speedwell

Kindness Marsh Mallow

Kindness and worth Czar Violet

Kindness, humility Elderberry Blossom

Lasting pleasure Everlasting Pea

Iris

Dog Rose

Patience Dock · Oxeye

Pity Pine

Pleasantry Gentle Balm

Pleasure and pain Dog Rose

Reciprocity, I share your sentiments
  China Aster (double)

Relieve my anxiety Christmas Rose

Safety Traveller's Joy

Self-sacrifice Andromeda

Sensitivity Mimosa

Sincerity Garden Chervil

Solace in adversity Evergreen Thorn

Strength Cedar Tree

Strength, union Bindweed

Strong friendship American Ivy

Succour, protection Juniper

Sympathy Lemon Balm · Thrift

Thankfulness, gratitude Agrimony

The variety of your conversation delights me Clarkia

Thoughtfulness Brier

Truth, platonic love Bittersweet

Unanimity Phlox

Agrimony

Unchanging friendship Thuja

Union, agreement Straw (whole)

Unpretentiousness Pasque Flower

Warmth Cactus

Warmth of feeling Peppermint

Warmth of sentiment Spearmint

You please all Currants (branch of)

Your friendship is agreeable to me

Glycine

Your presence softens my pains Milkvetch

Your presence soothes me Petunia

Your qualities, like your charms,

are unequalled Peach

Your qualities surpass your charms

Mignionette

Evil Deeds
& Dishonourable
Actions

Animosity St John's Wort

Antipathy Carnation (purple)

Baseness Dodder of Thyme

Contempt Sultan (yellow)

Crime Tamarisk

Dangerous pleasures Tuberose

Deformity Begonia

Disdain Carnation (yellow) · Rue

Disgust Frog Orchid

Ensnare Dragon Plant

Evil, you will cause my death Hemlock

Tamarisk

Lobelia

Hate for mankind Wolfsbane

Hatred, disdain Fire Lily

Horror Creeping Cereus · Dragonwort

· Mandrake · Snake Plant

I declare against you Purple Milkvetch

I declare war on you Wild Tansy

Ill-natured beauty Citron

Injustice Chestnut · Hop

Malevolence Lobelia

Meanness Cuscuta

Misanthropy Fuller's Thistle

Persecution Snake's Head Fritillary

Retaliation Scotch Thistle

Revenge Birdsfoot Trefoil

Suspicion Mushroom

Voluptuousness Saffron

Voraciousness Lupin

War Milfoil · York and Lancaster Rose

Your looks freeze me Ice Plant

———✦✦✦✦✦✦✦———

Bravery, Humanity
& The Pursuit
of Intellectual
Excellence

———✦✦✦✦✦✦✦———

A proud spirit Gloxinia

Adroitness Spider Orchid

Ambition Hollyhock · Mountain Laurel

Ambition in women Double-Blossomed
  Nasturtium

Boldness Pink

Boldness, audacity Larch

Bravery Oak Leaves

Bravery, humanity French Willow

Chivalry Daffodil (yellow) · Monkshood

Completeness Strawberry Leaves

Monkshood

Tulip

Confidence Liverwort · Polyanthus (lilac)

Courage Borage · Black Poplar

Courage, strength Garlic

Distinction Cardinal Flower

Education Cherry Tree

Eloquence Lagerstroemia

Endurance Cattail

Energy Salvia (red)

Energy in adversity Chamomile

Expertise Germander Speedwell

Fame Trumpet Flower · Tulip

Foresight Holly · Strawberry Blossom

Fortitude Fennel

Freedom Water Willow

Gallantry Sweet William

Genius Plane Tree

Glory Daphne · Laurel

I surmount all difficulties Mistletoe

Independence White Oak · Wild Plum

Industry Clover (red)

Holly

Rudbeckia

Inspiration Angelica

Intellect Walnut

Intellectual excellence and splendour

  Venice Sumach

Justice Rudbeckia

Justice shall be done Coltsfoot

Liberty Live Oak

Majesty Rose (first of summer)

Order Fir Cone

Patriotism American Elm · Nasturtium

Power, majesty Crown Imperial

Prepared to fight! Gladiola

Promptness Ten-week Stock

Quick sightedness Hawkweed

Rarity Ivy (white)

Reason Goat's Rue

Resistance Tansy

Resolved to win Columbine (purple)

Rigour Lantana

Success crown your wishes Coronella

Superior merit Moss Rose

Utility Flax (dried)

Worth beyond beauty Sweet Alyssum

Arrivals & Departures

Wormwood (detail)

Absence Wormwood

Awaiting Forsythia

Delay Eupatorium

Departure Sweet Pea

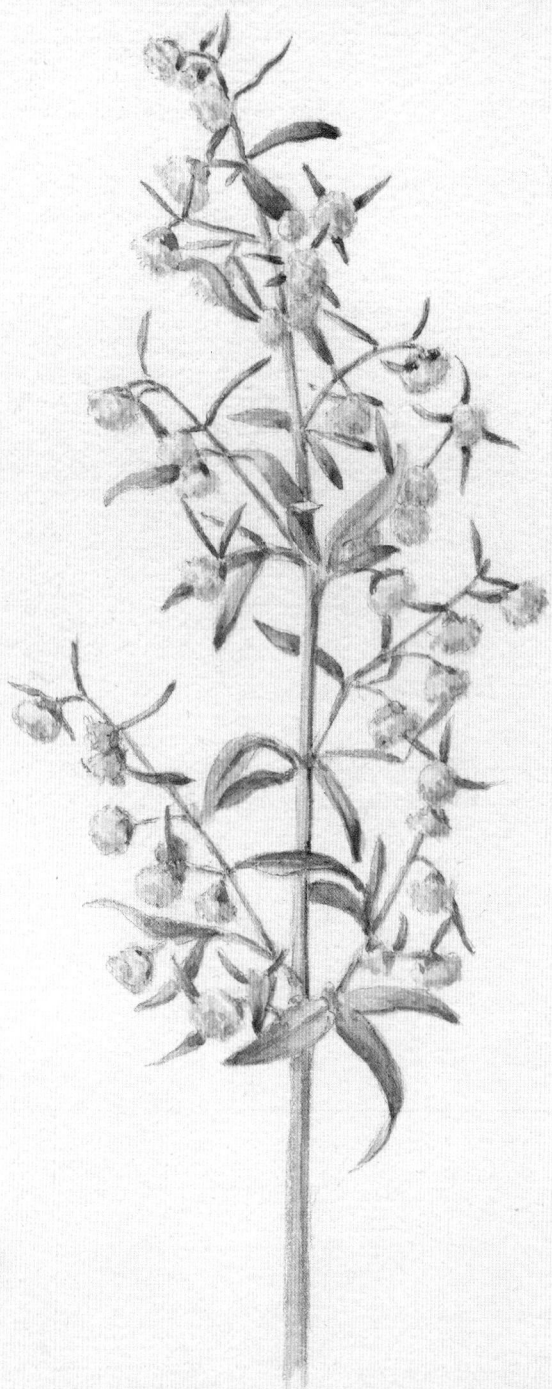

Wormwood

Farewell Michaelmas Daisy

Goodbye Cyclamen

Greeting Holly berry

Recall Geranium (sliver-leaf)

Welcome Wisteria

Welcome me Old Pink Daily Rose

Welcome to a stranger Eurasian Starwort

Birth, Death, Grief & Sorrow

Agitation Sainfoin

Anxious and trembling Columbine (red)

Bashful shame Rose (deep red)

Birth Dittany of Crete

Death is preferable to loss of innocence
  Rose (dried white)

Death, mourning Cypress

Dejection Lichen

Despondency Humble Plant

Grief Harebell · Marigold

Harebell

Meadow Saffron

Hopelessness Love-lies-bleeding

I die if neglected Laurestina

I shall die tomorrow Gum Cistus

I shall not survive you

   Black Mulberry Tree

Lamentation Aspen Tree

Melancholy, sadness

   Geranium (dark purple)

Mourning Willow

My best days are passed Meadow Saffron

My regrets follow you to the grave

   Asphodel

Refusal, sorry I can't be with you

   Carnation (striped)

Regret Verbena (purple)

Remorse Raspberry

Rustic oracle Dandelion

Separation Trumpet Vine

Sorrow Hyacinth (purple)

Sorrow and sadness Yew

Suffering Indian Plum

Symbol of life Acorn

Tears Helenium

Widowhood Sweet Scabiosa

Peace, Prudence & Virtuous Behaviour

Calm repose Buckbean

Character, loveliness Camellia

Discipline Wax Myrtle

Discretion Solomon's Seal

Domestic industry Flax · Houseleek

Domestic virtue Sage

Frugality Chicory

Gentility Corn Cockle

Good taste Daffodil (single)

Gratitude Bellflower (small, white)

Health Iceland Moss

Corn Cockle

Hawthorn

Hope Almond · Hawthorn · Snowdrop

Hope for better days Marianthus

Hope in adversity Spruce Pine

Incorruptible Cedar of Lebanon

Integrity Gentian

Meditation Chinese Lantern

Noble courage Edelweiss

Peace Olive

Peace, reconciliation Hazel

Peace, refinement Gardenia

Perseverance Canary Grass

    • Ground Laurel • Swamp Magnolia

Prudence Mountain Ash

    • Service Tree

Repose Dwarf Morning Glory (blue)

Reserve Maple Tree

Restraint Azalea

Secrecy Maidenhair Fern

Silence Belladonna • Oriental Poppy

Sleep Poppy (white)

Solitude, privacy Heath

Belladonna

Stoicism Box Tree

Talent Pink (white)

Taste Fuchsia

The fine arts (admiration of) Acanthus

Tranquility Stonecrop

Truth Chrysanthemum (white)

Truth, eloquence Lotus

Unobtrusive loveliness Hyacinth (white)

Unpretending excellence

   Japanese Camellia (red)

Wisdom White Mulberry Tree • Salvia (blue)

Thoughts, Memories & Times Past

Always remembered Everlasting Flower

Am I forgotten? Satin Flower

Forgetfulness Moonwort

Lasting memories Spindleberry

Memory Syringa

Painful memories Adonis

Pleasant memories Periwinkle (white)

Remembrance Rosemary · Pheasant's Eye

Remembrance, consolation Poppy (red)

Poppy (red)

Reverie Flowering Fern

Think of me Cedar Leaf · Clover (white)

Thoughts of absent friends Zinnia

Time Fir

Time White Poplar

Unhappy memories Milkweed

You occupy my thoughts Pansy

· Violet (purple)

Index of Flowers

Index of Meanings